# MATH WORKBOOK

# ADDITION

1)
$$\begin{array}{r} 234 \\ +\ 175 \\ \hline \end{array}$$
..........................................

2)
$$\begin{array}{r} 229 \\ +\ 130 \\ \hline \end{array}$$
..........................................

3)
$$\begin{array}{r} 271 \\ +\ 119 \\ \hline \end{array}$$
..........................................

4)
$$\begin{array}{r} 147 \\ +\ 259 \\ \hline \end{array}$$
..........................................

5)
$$\begin{array}{r} 125 \\ +\ 266 \\ \hline \end{array}$$
..........................................

6)
$$\begin{array}{r} 129 \\ +\ 146 \\ \hline \end{array}$$
..........................................

7)
$$\begin{array}{r} 249 \\ +\ 191 \\ \hline \end{array}$$
..........................................

8)
$$\begin{array}{r} 277 \\ +\ 117 \\ \hline \end{array}$$
..........................................

9)
$$\begin{array}{r} 154 \\ +\ 266 \\ \hline \end{array}$$
..........................................

10)
$$\begin{array}{r} 285 \\ +\ 240 \\ \hline \end{array}$$
..........................................

11)
$$\begin{array}{r} 286 \\ +\ 235 \\ \hline \end{array}$$
..........................................

12)
$$\begin{array}{r} 119 \\ +\ 247 \\ \hline \end{array}$$
..........................................

13)
$$\begin{array}{r} 261 \\ +\ 251 \\ \hline \end{array}$$
..........................................

14)
$$\begin{array}{r} 292 \\ +\ 157 \\ \hline \end{array}$$
..........................................

15)
$$\begin{array}{r} 289 \\ +\ 245 \\ \hline \end{array}$$
..........................................

1) $\begin{array}{r} 125 \\ +\,279 \\ \hline \end{array}$ ..............................

2) $\begin{array}{r} 164 \\ +\,293 \\ \hline \end{array}$ ..............................

3) $\begin{array}{r} 222 \\ +\,161 \\ \hline \end{array}$ ..............................

4) $\begin{array}{r} 130 \\ +\,278 \\ \hline \end{array}$ ..............................

5) $\begin{array}{r} 123 \\ +\,165 \\ \hline \end{array}$ ..............................

6) $\begin{array}{r} 299 \\ +\,156 \\ \hline \end{array}$ ..............................

7) $\begin{array}{r} 282 \\ +\,169 \\ \hline \end{array}$ ..............................

8) $\begin{array}{r} 120 \\ +\,127 \\ \hline \end{array}$ ..............................

9) $\begin{array}{r} 289 \\ +\,271 \\ \hline \end{array}$ ..............................

10) $\begin{array}{r} 180 \\ +\,102 \\ \hline \end{array}$ ..............................

11) $\begin{array}{r} 173 \\ +\,121 \\ \hline \end{array}$ ..............................

12) $\begin{array}{r} 162 \\ +\,252 \\ \hline \end{array}$ ..............................

13) $\begin{array}{r} 128 \\ +\,107 \\ \hline \end{array}$ ..............................

14) $\begin{array}{r} 199 \\ +\,262 \\ \hline \end{array}$ ..............................

15) $\begin{array}{r} 183 \\ +\,290 \\ \hline \end{array}$ ..............................

1) $\begin{array}{r} 193 \\ +\,127 \\ \hline \end{array}$ ..........

2) $\begin{array}{r} 187 \\ +\,283 \\ \hline \end{array}$ ..........

3) $\begin{array}{r} 174 \\ +\,261 \\ \hline \end{array}$ ..........

4) $\begin{array}{r} 223 \\ +\,186 \\ \hline \end{array}$ ..........

5) $\begin{array}{r} 217 \\ +\,106 \\ \hline \end{array}$ ..........

6) $\begin{array}{r} 104 \\ +\,217 \\ \hline \end{array}$ ..........

7) $\begin{array}{r} 212 \\ +\,265 \\ \hline \end{array}$ ..........

8) $\begin{array}{r} 215 \\ +\,262 \\ \hline \end{array}$ ..........

9) $\begin{array}{r} 167 \\ +\,194 \\ \hline \end{array}$ ..........

10) $\begin{array}{r} 133 \\ +\,273 \\ \hline \end{array}$ ..........

11) $\begin{array}{r} 130 \\ +\,256 \\ \hline \end{array}$ ..........

12) $\begin{array}{r} 215 \\ +\,233 \\ \hline \end{array}$ ..........

13) $\begin{array}{r} 148 \\ +\,200 \\ \hline \end{array}$ ..........

14) $\begin{array}{r} 225 \\ +\,204 \\ \hline \end{array}$ ..........

15) $\begin{array}{r} 295 \\ +\,167 \\ \hline \end{array}$ ..........

1) $155 + 115$ = ..........

2) $199 + 252$ = ..........

3) $223 + 167$ = ..........

4) $184 + 269$ = ..........

5) $150 + 210$ = ..........

6) $238 + 212$ = ..........

7) $254 + 235$ = ..........

8) $274 + 135$ = ..........

9) $119 + 190$ = ..........

10) $149 + 213$ = ..........

11) $286 + 177$ = ..........

12) $200 + 130$ = ..........

13) $234 + 250$ = ..........

14) $188 + 185$ = ..........

15) $104 + 104$ = ..........

1) $\begin{array}{r} 168 \\ +\,158 \\ \hline \end{array}$ ..........

2) $\begin{array}{r} 144 \\ +\,136 \\ \hline \end{array}$ ..........

3) $\begin{array}{r} 294 \\ +\,250 \\ \hline \end{array}$ ..........

4) $\begin{array}{r} 293 \\ +\,235 \\ \hline \end{array}$ ..........

5) $\begin{array}{r} 249 \\ +\,246 \\ \hline \end{array}$ ..........

6) $\begin{array}{r} 278 \\ +\,215 \\ \hline \end{array}$ ..........

7) $\begin{array}{r} 170 \\ +\,118 \\ \hline \end{array}$ ..........

8) $\begin{array}{r} 273 \\ +\,278 \\ \hline \end{array}$ ..........

9) $\begin{array}{r} 299 \\ +\,265 \\ \hline \end{array}$ ..........

10) $\begin{array}{r} 146 \\ +\,263 \\ \hline \end{array}$ ..........

11) $\begin{array}{r} 266 \\ +\,159 \\ \hline \end{array}$ ..........

12) $\begin{array}{r} 253 \\ +\,241 \\ \hline \end{array}$ ..........

13) $\begin{array}{r} 189 \\ +\,109 \\ \hline \end{array}$ ..........

14) $\begin{array}{r} 167 \\ +\,118 \\ \hline \end{array}$ ..........

15) $\begin{array}{r} 149 \\ +\,203 \\ \hline \end{array}$ ..........

1) $$\begin{array}{r} 567 \\ +\ 627 \\ \hline \end{array}$$

......................................................

2) $$\begin{array}{r} 449 \\ +\ 585 \\ \hline \end{array}$$

......................................................

3) $$\begin{array}{r} 213 \\ +\ 525 \\ \hline \end{array}$$

......................................................

4) $$\begin{array}{r} 768 \\ +\ 538 \\ \hline \end{array}$$

......................................................

5) $$\begin{array}{r} 439 \\ +\ 108 \\ \hline \end{array}$$

......................................................

6) $$\begin{array}{r} 736 \\ +\ 673 \\ \hline \end{array}$$

......................................................

7) $$\begin{array}{r} 367 \\ +\ 420 \\ \hline \end{array}$$

......................................................

8) $$\begin{array}{r} 178 \\ +\ 531 \\ \hline \end{array}$$

......................................................

9) $$\begin{array}{r} 422 \\ +\ 469 \\ \hline \end{array}$$

......................................................

10) $$\begin{array}{r} 394 \\ +\ 715 \\ \hline \end{array}$$

......................................................

11) $$\begin{array}{r} 765 \\ +\ 386 \\ \hline \end{array}$$

......................................................

12) $$\begin{array}{r} 406 \\ +\ 535 \\ \hline \end{array}$$

......................................................

13) $$\begin{array}{r} 431 \\ +\ 147 \\ \hline \end{array}$$

......................................................

14) $$\begin{array}{r} 796 \\ +\ 559 \\ \hline \end{array}$$

......................................................

15) $$\begin{array}{r} 461 \\ +\ 543 \\ \hline \end{array}$$

......................................................

1)
$$\begin{array}{r} 2\,0\,3 \\ +\ 6\,9\,7 \\ \hline \end{array}$$
........................

2)
$$\begin{array}{r} 2\,3\,7 \\ +\ 7\,9\,5 \\ \hline \end{array}$$
........................

3)
$$\begin{array}{r} 4\,6\,9 \\ +\ 3\,9\,7 \\ \hline \end{array}$$
........................

4)
$$\begin{array}{r} 7\,6\,6 \\ +\ 7\,0\,1 \\ \hline \end{array}$$
........................

5)
$$\begin{array}{r} 1\,9\,0 \\ +\ 3\,7\,4 \\ \hline \end{array}$$
........................

6)
$$\begin{array}{r} 2\,8\,1 \\ +\ 3\,2\,0 \\ \hline \end{array}$$
........................

7)
$$\begin{array}{r} 7\,1\,4 \\ +\ 2\,5\,4 \\ \hline \end{array}$$
........................

8)
$$\begin{array}{r} 1\,6\,6 \\ +\ 5\,3\,0 \\ \hline \end{array}$$
........................

9)
$$\begin{array}{r} 3\,0\,9 \\ +\ 1\,7\,8 \\ \hline \end{array}$$
........................

10)
$$\begin{array}{r} 6\,9\,5 \\ +\ 2\,5\,2 \\ \hline \end{array}$$
........................

11)
$$\begin{array}{r} 2\,1\,3 \\ +\ 7\,5\,6 \\ \hline \end{array}$$
........................

12)
$$\begin{array}{r} 5\,1\,2 \\ +\ 4\,2\,2 \\ \hline \end{array}$$
........................

13)
$$\begin{array}{r} 1\,8\,2 \\ +\ 6\,8\,3 \\ \hline \end{array}$$
........................

14)
$$\begin{array}{r} 6\,9\,9 \\ +\ 7\,4\,6 \\ \hline \end{array}$$
........................

15)
$$\begin{array}{r} 2\,6\,2 \\ +\ 1\,6\,2 \\ \hline \end{array}$$
........................

1) $\begin{array}{r} 576 \\ +\ 734 \\ \hline \end{array}$ ..........

2) $\begin{array}{r} 369 \\ +\ 458 \\ \hline \end{array}$ ..........

3) $\begin{array}{r} 503 \\ +\ 444 \\ \hline \end{array}$ ..........

4) $\begin{array}{r} 498 \\ +\ 696 \\ \hline \end{array}$ ..........

5) $\begin{array}{r} 273 \\ +\ 380 \\ \hline \end{array}$ ..........

6) $\begin{array}{r} 247 \\ +\ 285 \\ \hline \end{array}$ ..........

7) $\begin{array}{r} 615 \\ +\ 713 \\ \hline \end{array}$ ..........

8) $\begin{array}{r} 587 \\ +\ 498 \\ \hline \end{array}$ ..........

9) $\begin{array}{r} 120 \\ +\ 539 \\ \hline \end{array}$ ..........

10) $\begin{array}{r} 128 \\ +\ 185 \\ \hline \end{array}$ ..........

11) $\begin{array}{r} 486 \\ +\ 339 \\ \hline \end{array}$ ..........

12) $\begin{array}{r} 673 \\ +\ 726 \\ \hline \end{array}$ ..........

13) $\begin{array}{r} 716 \\ +\ 672 \\ \hline \end{array}$ ..........

14) $\begin{array}{r} 545 \\ +\ 186 \\ \hline \end{array}$ ..........

15) $\begin{array}{r} 549 \\ +\ 361 \\ \hline \end{array}$ ..........

1) 764 + 143 = ..............

2) 651 + 305 = ..............

3) 548 + 670 = ..............

4) 284 + 599 = ..............

5) 295 + 231 = ..............

6) 337 + 161 = ..............

7) 443 + 542 = ..............

8) 631 + 700 = ..............

9) 412 + 247 = ..............

10) 712 + 189 = ..............

11) 118 + 154 = ..............

12) 110 + 407 = ..............

13) 638 + 168 = ..............

14) 531 + 779 = ..............

15) 411 + 395 = ..............

1) 688 + 788 = ..........

2) 647 + 728 = ..........

3) 618 + 278 = ..........

4) 552 + 697 = ..........

5) 472 + 163 = ..........

6) 710 + 134 = ..........

7) 344 + 467 = ..........

8) 480 + 239 = ..........

9) 304 + 288 = ..........

10) 701 + 443 = ..........

11) 607 + 750 = ..........

12) 597 + 590 = ..........

13) 102 + 106 = ..........

14) 531 + 246 = ..........

15) 321 + 182 = ..........

1)

| | 2 | 2 | 4 | 0 |
|---|---|---|---|---|
| + | 1 | 9 | 3 | 1 |

..............................................................

2)

| | 1 | 1 | 0 | 0 |
|---|---|---|---|---|
| + | 2 | 8 | 5 | 5 |

..............................................................

3)

| | 1 | 8 | 6 | 6 |
|---|---|---|---|---|
| + | 1 | 8 | 5 | 1 |

..............................................................

4)

| | 1 | 8 | 7 | 1 |
|---|---|---|---|---|
| + | 1 | 3 | 5 | 1 |

..............................................................

5)

| | 1 | 7 | 0 | 5 |
|---|---|---|---|---|
| + | 2 | 5 | 1 | 8 |

..............................................................

6)

| | 1 | 6 | 8 | 1 |
|---|---|---|---|---|
| + | 2 | 9 | 6 | 5 |

..............................................................

7)

| | 2 | 3 | 9 | 6 |
|---|---|---|---|---|
| + | 2 | 1 | 9 | 8 |

..............................................................

8)

| | 1 | 7 | 0 | 8 |
|---|---|---|---|---|
| + | 1 | 4 | 6 | 0 |

..............................................................

9)

| | 2 | 1 | 1 | 4 |
|---|---|---|---|---|
| + | 1 | 8 | 8 | 7 |

..............................................................

10)

| | 1 | 6 | 1 | 5 |
|---|---|---|---|---|
| + | 1 | 8 | 9 | 3 |

..............................................................

1) $\begin{array}{r} 1\,655 \\ +\ 2\,766 \\ \hline \end{array}$

..................................................

2) $\begin{array}{r} 2\,170 \\ +\ 2\,092 \\ \hline \end{array}$

..................................................

3) $\begin{array}{r} 1\,893 \\ +\ 1\,274 \\ \hline \end{array}$

..................................................

4) $\begin{array}{r} 1\,807 \\ +\ 1\,872 \\ \hline \end{array}$

..................................................

5) $\begin{array}{r} 2\,718 \\ +\ 1\,442 \\ \hline \end{array}$

..................................................

6) $\begin{array}{r} 2\,286 \\ +\ 1\,450 \\ \hline \end{array}$

..................................................

7) $\begin{array}{r} 1\,679 \\ +\ 1\,252 \\ \hline \end{array}$

..................................................

8) $\begin{array}{r} 2\,381 \\ +\ 1\,641 \\ \hline \end{array}$

..................................................

9) $\begin{array}{r} 1\,197 \\ +\ 2\,125 \\ \hline \end{array}$

..................................................

10) $\begin{array}{r} 1\,361 \\ +\ 2\,054 \\ \hline \end{array}$

..................................................

1)

$$\begin{array}{r} 2\ 461 \\ +\ 1\ 840 \\ \hline \end{array}$$

..................................................

2)

$$\begin{array}{r} 2\ 397 \\ +\ 2\ 085 \\ \hline \end{array}$$

..................................................

3)

$$\begin{array}{r} 1\ 863 \\ +\ 1\ 915 \\ \hline \end{array}$$

..................................................

4)

$$\begin{array}{r} 1\ 796 \\ +\ 1\ 304 \\ \hline \end{array}$$

..................................................

5)

$$\begin{array}{r} 1\ 305 \\ +\ 1\ 434 \\ \hline \end{array}$$

..................................................

6)

$$\begin{array}{r} 2\ 654 \\ +\ 1\ 241 \\ \hline \end{array}$$

..................................................

7)

$$\begin{array}{r} 2\ 169 \\ +\ 1\ 173 \\ \hline \end{array}$$

..................................................

8)

$$\begin{array}{r} 2\ 608 \\ +\ 1\ 380 \\ \hline \end{array}$$

..................................................

9)

$$\begin{array}{r} 1\ 132 \\ +\ 1\ 105 \\ \hline \end{array}$$

..................................................

10)

$$\begin{array}{r} 1\ 183 \\ +\ 1\ 748 \\ \hline \end{array}$$

..................................................

1)
$$\begin{array}{r} 2\ 701 \\ +\ 1\ 271 \\ \hline \end{array}$$

..........

2)
$$\begin{array}{r} 1\ 407 \\ +\ 1\ 801 \\ \hline \end{array}$$

..........

3)
$$\begin{array}{r} 1\ 981 \\ +\ 1\ 446 \\ \hline \end{array}$$

..........

4)
$$\begin{array}{r} 2\ 053 \\ +\ 2\ 595 \\ \hline \end{array}$$

..........

5)
$$\begin{array}{r} 2\ 389 \\ +\ 1\ 466 \\ \hline \end{array}$$

..........

6)
$$\begin{array}{r} 2\ 846 \\ +\ 2\ 893 \\ \hline \end{array}$$

..........

7)
$$\begin{array}{r} 1\ 820 \\ +\ 2\ 927 \\ \hline \end{array}$$

..........

8)
$$\begin{array}{r} 2\ 400 \\ +\ 2\ 718 \\ \hline \end{array}$$

..........

9)
$$\begin{array}{r} 2\ 141 \\ +\ 2\ 633 \\ \hline \end{array}$$

..........

10)
$$\begin{array}{r} 2\ 013 \\ +\ 2\ 700 \\ \hline \end{array}$$

..........

1)
$$\begin{array}{r} 1\,954 \\ +\ 1\,247 \\ \hline \end{array}$$
..............................................................

2)
$$\begin{array}{r} 2\,291 \\ +\ 2\,841 \\ \hline \end{array}$$
..............................................................

3)
$$\begin{array}{r} 1\,733 \\ +\ 2\,382 \\ \hline \end{array}$$
..............................................................

4)
$$\begin{array}{r} 2\,834 \\ +\ 1\,368 \\ \hline \end{array}$$
..............................................................

5)
$$\begin{array}{r} 2\,192 \\ +\ 2\,161 \\ \hline \end{array}$$
..............................................................

6)
$$\begin{array}{r} 2\,327 \\ +\ 1\,537 \\ \hline \end{array}$$
..............................................................

7)
$$\begin{array}{r} 1\,715 \\ +\ 2\,717 \\ \hline \end{array}$$
..............................................................

8)
$$\begin{array}{r} 1\,902 \\ +\ 2\,859 \\ \hline \end{array}$$
..............................................................

9)
$$\begin{array}{r} 2\,445 \\ +\ 2\,743 \\ \hline \end{array}$$
..............................................................

10)
$$\begin{array}{r} 1\,499 \\ +\ 2\,943 \\ \hline \end{array}$$
..............................................................

1) $2\,277 + 2\,555$ = ..........

2) $1\,171 + 1\,871$ = ..........

3) $2\,512 + 2\,599$ = ..........

4) $1\,602 + 1\,979$ = ..........

5) $2\,194 + 1\,018$ = ..........

6) $2\,884 + 1\,772$ = ..........

7) $1\,441 + 2\,852$ = ..........

8) $1\,373 + 2\,760$ = ..........

9) $2\,753 + 1\,635$ = ..........

10) $1\,816 + 2\,307$ = ..........

1)

$$\begin{array}{r} 2\ 463 \\ +\ 2\ 145 \\ \hline \end{array}$$

..........

2)

$$\begin{array}{r} 1\ 733 \\ +\ 2\ 407 \\ \hline \end{array}$$

..........

3)

$$\begin{array}{r} 2\ 884 \\ +\ 2\ 773 \\ \hline \end{array}$$

..........

4)

$$\begin{array}{r} 1\ 366 \\ +\ 2\ 844 \\ \hline \end{array}$$

..........

5)

$$\begin{array}{r} 1\ 204 \\ +\ 2\ 303 \\ \hline \end{array}$$

..........

6)

$$\begin{array}{r} 2\ 257 \\ +\ 1\ 466 \\ \hline \end{array}$$

..........

7)

$$\begin{array}{r} 1\ 548 \\ +\ 1\ 481 \\ \hline \end{array}$$

..........

8)

$$\begin{array}{r} 1\ 060 \\ +\ 1\ 432 \\ \hline \end{array}$$

..........

9)

$$\begin{array}{r} 2\ 887 \\ +\ 2\ 683 \\ \hline \end{array}$$

..........

10)

$$\begin{array}{r} 1\ 940 \\ +\ 1\ 408 \\ \hline \end{array}$$

..........

1) 
$$\begin{array}{r} 1\,887 \\ +\ 1\,623 \\ \hline \end{array}$$
..................................................

2) 
$$\begin{array}{r} 2\,306 \\ +\ 2\,693 \\ \hline \end{array}$$
..................................................

3) 
$$\begin{array}{r} 1\,555 \\ +\ 1\,738 \\ \hline \end{array}$$
..................................................

4) 
$$\begin{array}{r} 1\,040 \\ +\ 2\,034 \\ \hline \end{array}$$
..................................................

5) 
$$\begin{array}{r} 2\,968 \\ +\ 1\,600 \\ \hline \end{array}$$
..................................................

6) 
$$\begin{array}{r} 2\,394 \\ +\ 2\,909 \\ \hline \end{array}$$
..................................................

7) 
$$\begin{array}{r} 1\,431 \\ +\ 2\,513 \\ \hline \end{array}$$
..................................................

8) 
$$\begin{array}{r} 1\,050 \\ +\ 2\,285 \\ \hline \end{array}$$
..................................................

9) 
$$\begin{array}{r} 1\,879 \\ +\ 1\,517 \\ \hline \end{array}$$
..................................................

10) 
$$\begin{array}{r} 2\,420 \\ +\ 1\,881 \\ \hline \end{array}$$
..................................................

1)
$$\begin{array}{r} 2\ 216 \\ +\ 2\ 674 \\ \hline \end{array}$$

..........

2)
$$\begin{array}{r} 1\ 286 \\ +\ 2\ 318 \\ \hline \end{array}$$

..........

3)
$$\begin{array}{r} 1\ 416 \\ +\ 1\ 634 \\ \hline \end{array}$$

..........

4)
$$\begin{array}{r} 2\ 160 \\ +\ 1\ 856 \\ \hline \end{array}$$

..........

5)
$$\begin{array}{r} 2\ 926 \\ +\ 2\ 321 \\ \hline \end{array}$$

..........

6)
$$\begin{array}{r} 1\ 393 \\ +\ 1\ 853 \\ \hline \end{array}$$

..........

7)
$$\begin{array}{r} 2\ 716 \\ +\ 1\ 340 \\ \hline \end{array}$$

..........

8)
$$\begin{array}{r} 2\ 014 \\ +\ 2\ 569 \\ \hline \end{array}$$

..........

9)
$$\begin{array}{r} 1\ 853 \\ +\ 1\ 360 \\ \hline \end{array}$$

..........

10)
$$\begin{array}{r} 1\ 279 \\ +\ 1\ 750 \\ \hline \end{array}$$

..........

1)
$$\begin{array}{r} 2\,466 \\ +\ 2\,635 \\ \hline \end{array}$$

.....................................

2)
$$\begin{array}{r} 2\,662 \\ +\ 2\,933 \\ \hline \end{array}$$

.....................................

3)
$$\begin{array}{r} 2\,384 \\ +\ 2\,720 \\ \hline \end{array}$$

.....................................

4)
$$\begin{array}{r} 2\,623 \\ +\ 2\,113 \\ \hline \end{array}$$

.....................................

5)
$$\begin{array}{r} 2\,097 \\ +\ 2\,416 \\ \hline \end{array}$$

.....................................

6)
$$\begin{array}{r} 1\,125 \\ +\ 1\,023 \\ \hline \end{array}$$

.....................................

7)
$$\begin{array}{r} 1\,591 \\ +\ 1\,099 \\ \hline \end{array}$$

.....................................

8)
$$\begin{array}{r} 1\,488 \\ +\ 2\,886 \\ \hline \end{array}$$

.....................................

9)
$$\begin{array}{r} 2\,313 \\ +\ 1\,699 \\ \hline \end{array}$$

.....................................

10)
$$\begin{array}{r} 1\,252 \\ +\ 1\,454 \\ \hline \end{array}$$

.....................................

# SUBTRACTION

1) $\begin{array}{r} 205 \\ -\,122 \\ \hline \end{array}$ ..............

2) $\begin{array}{r} 166 \\ -\,154 \\ \hline \end{array}$ ..............

3) $\begin{array}{r} 188 \\ -\,110 \\ \hline \end{array}$ ..............

4) $\begin{array}{r} 190 \\ -\,171 \\ \hline \end{array}$ ..............

5) $\begin{array}{r} 218 \\ -\,102 \\ \hline \end{array}$ ..............

6) $\begin{array}{r} 150 \\ -\,100 \\ \hline \end{array}$ ..............

7) $\begin{array}{r} 245 \\ -\,135 \\ \hline \end{array}$ ..............

8) $\begin{array}{r} 161 \\ -\,156 \\ \hline \end{array}$ ..............

9) $\begin{array}{r} 165 \\ -\,119 \\ \hline \end{array}$ ..............

10) $\begin{array}{r} 138 \\ -\,109 \\ \hline \end{array}$ ..............

11) $\begin{array}{r} 245 \\ -\,167 \\ \hline \end{array}$ ..............

12) $\begin{array}{r} 140 \\ -\,133 \\ \hline \end{array}$ ..............

13) $\begin{array}{r} 200 \\ -\,120 \\ \hline \end{array}$ ..............

14) $\begin{array}{r} 153 \\ -\,150 \\ \hline \end{array}$ ..............

15) $\begin{array}{r} 138 \\ -\,134 \\ \hline \end{array}$ ..............

1) $235 - 204$ = ..........

2) $113 - 109$ = ..........

3) $210 - 113$ = ..........

4) $143 - 125$ = ..........

5) $193 - 161$ = ..........

6) $136 - 134$ = ..........

7) $192 - 115$ = ..........

8) $158 - 144$ = ..........

9) $198 - 152$ = ..........

10) $233 - 136$ = ..........

11) $164 - 148$ = ..........

12) $246 - 176$ = ..........

13) $152 - 118$ = ..........

14) $158 - 129$ = ..........

15) $197 - 133$ = ..........

1) $116 - 116 =$ ..........

2) $233 - 153 =$ ..........

3) $227 - 130 =$ ..........

4) $191 - 147 =$ ..........

5) $180 - 133 =$ ..........

6) $121 - 117 =$ ..........

7) $184 - 126 =$ ..........

8) $159 - 143 =$ ..........

9) $217 - 210 =$ ..........

10) $142 - 135 =$ ..........

11) $128 - 114 =$ ..........

12) $194 - 137 =$ ..........

13) $243 - 169 =$ ..........

14) $155 - 101 =$ ..........

15) $116 - 113 =$ ..........

1) $\begin{array}{r} 109 \\ -\ 100 \\ \hline \end{array}$ ..........

2) $\begin{array}{r} 365 \\ -\ 165 \\ \hline \end{array}$ ..........

3) $\begin{array}{r} 192 \\ -\ 157 \\ \hline \end{array}$ ..........

4) $\begin{array}{r} 146 \\ -\ 109 \\ \hline \end{array}$ ..........

5) $\begin{array}{r} 207 \\ -\ 165 \\ \hline \end{array}$ ..........

6) $\begin{array}{r} 282 \\ -\ 129 \\ \hline \end{array}$ ..........

7) $\begin{array}{r} 463 \\ -\ 115 \\ \hline \end{array}$ ..........

8) $\begin{array}{r} 427 \\ -\ 100 \\ \hline \end{array}$ ..........

9) $\begin{array}{r} 191 \\ -\ 115 \\ \hline \end{array}$ ..........

10) $\begin{array}{r} 455 \\ -\ 195 \\ \hline \end{array}$ ..........

11) $\begin{array}{r} 123 \\ -\ 106 \\ \hline \end{array}$ ..........

12) $\begin{array}{r} 385 \\ -\ 117 \\ \hline \end{array}$ ..........

13) $\begin{array}{r} 340 \\ -\ 172 \\ \hline \end{array}$ ..........

14) $\begin{array}{r} 382 \\ -\ 154 \\ \hline \end{array}$ ..........

15) $\begin{array}{r} 339 \\ -\ 182 \\ \hline \end{array}$ ..........

1)
$$\begin{array}{r} 331 \\ -\ 172 \\ \hline \end{array}$$
...............................................

2)
$$\begin{array}{r} 417 \\ -\ 135 \\ \hline \end{array}$$
...............................................

3)
$$\begin{array}{r} 152 \\ -\ 127 \\ \hline \end{array}$$
...............................................

4)
$$\begin{array}{r} 344 \\ -\ 135 \\ \hline \end{array}$$
...............................................

5)
$$\begin{array}{r} 148 \\ -\ 127 \\ \hline \end{array}$$
...............................................

6)
$$\begin{array}{r} 397 \\ -\ 145 \\ \hline \end{array}$$
...............................................

7)
$$\begin{array}{r} 319 \\ -\ 225 \\ \hline \end{array}$$
...............................................

8)
$$\begin{array}{r} 399 \\ -\ 237 \\ \hline \end{array}$$
...............................................

9)
$$\begin{array}{r} 401 \\ -\ 190 \\ \hline \end{array}$$
...............................................

10)
$$\begin{array}{r} 389 \\ -\ 155 \\ \hline \end{array}$$
...............................................

11)
$$\begin{array}{r} 261 \\ -\ 163 \\ \hline \end{array}$$
...............................................

12)
$$\begin{array}{r} 280 \\ -\ 114 \\ \hline \end{array}$$
...............................................

13)
$$\begin{array}{r} 340 \\ -\ 248 \\ \hline \end{array}$$
...............................................

14)
$$\begin{array}{r} 308 \\ -\ 133 \\ \hline \end{array}$$
...............................................

15)
$$\begin{array}{r} 454 \\ -\ 113 \\ \hline \end{array}$$
...............................................

1) 468 − 170 = ..........

2) 449 − 130 = ..........

3) 135 − 128 = ..........

4) 408 − 111 = ..........

5) 420 − 171 = ..........

6) 162 − 102 = ..........

7) 381 − 195 = ..........

8) 159 − 100 = ..........

9) 257 − 152 = ..........

10) 114 − 112 = ..........

11) 204 − 179 = ..........

12) 222 − 199 = ..........

13) 399 − 100 = ..........

14) 351 − 120 = ..........

15) 381 − 212 = ..........

1) $\begin{array}{r} 488 \\ -\ 131 \\ \hline \end{array}$ ............

2) $\begin{array}{r} 295 \\ -\ 148 \\ \hline \end{array}$ ............

3) $\begin{array}{r} 106 \\ -\ 101 \\ \hline \end{array}$ ............

4) $\begin{array}{r} 289 \\ -\ 192 \\ \hline \end{array}$ ............

5) $\begin{array}{r} 113 \\ -\ 100 \\ \hline \end{array}$ ............

6) $\begin{array}{r} 165 \\ -\ 111 \\ \hline \end{array}$ ............

7) $\begin{array}{r} 124 \\ -\ 111 \\ \hline \end{array}$ ............

8) $\begin{array}{r} 135 \\ -\ 106 \\ \hline \end{array}$ ............

9) $\begin{array}{r} 355 \\ -\ 156 \\ \hline \end{array}$ ............

10) $\begin{array}{r} 122 \\ -\ 120 \\ \hline \end{array}$ ............

11) $\begin{array}{r} 253 \\ -\ 172 \\ \hline \end{array}$ ............

12) $\begin{array}{r} 450 \\ -\ 243 \\ \hline \end{array}$ ............

13) $\begin{array}{r} 353 \\ -\ 153 \\ \hline \end{array}$ ............

14) $\begin{array}{r} 251 \\ -\ 182 \\ \hline \end{array}$ ............

15) $\begin{array}{r} 398 \\ -\ 242 \\ \hline \end{array}$ ............

1) $332 - 317 =$ ..........

2) $438 - 275 =$ ..........

3) $263 - 174 =$ ..........

4) $241 - 178 =$ ..........

5) $263 - 222 =$ ..........

6) $242 - 173 =$ ..........

7) $672 - 177 =$ ..........

8) $592 - 151 =$ ..........

9) $128 - 100 =$ ..........

10) $359 - 238 =$ ..........

11) $702 - 595 =$ ..........

12) $489 - 326 =$ ..........

13) $649 - 513 =$ ..........

14) $663 - 243 =$ ..........

15) $508 - 229 =$ ..........

1) $\begin{array}{r} 821 \\ -\ 814 \\ \hline \end{array}$
.................................

2) $\begin{array}{r} 514 \\ -\ 213 \\ \hline \end{array}$
.................................

3) $\begin{array}{r} 489 \\ -\ 452 \\ \hline \end{array}$
.................................

4) $\begin{array}{r} 517 \\ -\ 125 \\ \hline \end{array}$
.................................

5) $\begin{array}{r} 533 \\ -\ 448 \\ \hline \end{array}$
.................................

6) $\begin{array}{r} 475 \\ -\ 268 \\ \hline \end{array}$
.................................

7) $\begin{array}{r} 773 \\ -\ 660 \\ \hline \end{array}$
.................................

8) $\begin{array}{r} 462 \\ -\ 372 \\ \hline \end{array}$
.................................

9) $\begin{array}{r} 837 \\ -\ 133 \\ \hline \end{array}$
.................................

10) $\begin{array}{r} 810 \\ -\ 625 \\ \hline \end{array}$
.................................

11) $\begin{array}{r} 325 \\ -\ 211 \\ \hline \end{array}$
.................................

12) $\begin{array}{r} 792 \\ -\ 468 \\ \hline \end{array}$
.................................

13) $\begin{array}{r} 508 \\ -\ 281 \\ \hline \end{array}$
.................................

14) $\begin{array}{r} 341 \\ -\ 288 \\ \hline \end{array}$
.................................

15) $\begin{array}{r} 448 \\ -\ 405 \\ \hline \end{array}$
.................................

1) $\begin{array}{r} 627 \\ -\ 532 \\ \hline \end{array}$ ..........

2) $\begin{array}{r} 821 \\ -\ 657 \\ \hline \end{array}$ ..........

3) $\begin{array}{r} 210 \\ -\ 181 \\ \hline \end{array}$ ..........

4) $\begin{array}{r} 397 \\ -\ 267 \\ \hline \end{array}$ ..........

5) $\begin{array}{r} 827 \\ -\ 374 \\ \hline \end{array}$ ..........

6) $\begin{array}{r} 426 \\ -\ 141 \\ \hline \end{array}$ ..........

7) $\begin{array}{r} 889 \\ -\ 703 \\ \hline \end{array}$ ..........

8) $\begin{array}{r} 134 \\ -\ 121 \\ \hline \end{array}$ ..........

9) $\begin{array}{r} 692 \\ -\ 470 \\ \hline \end{array}$ ..........

10) $\begin{array}{r} 631 \\ -\ 348 \\ \hline \end{array}$ ..........

11) $\begin{array}{r} 178 \\ -\ 148 \\ \hline \end{array}$ ..........

12) $\begin{array}{r} 835 \\ -\ 134 \\ \hline \end{array}$ ..........

13) $\begin{array}{r} 653 \\ -\ 413 \\ \hline \end{array}$ ..........

14) $\begin{array}{r} 853 \\ -\ 116 \\ \hline \end{array}$ ..........

15) $\begin{array}{r} 482 \\ -\ 196 \\ \hline \end{array}$ ..........

1)
$$\begin{array}{r} 1\;885 \\ -\;1\;517 \\ \hline \end{array}$$
..............................

2)
$$\begin{array}{r} 1\;626 \\ -\;1\;174 \\ \hline \end{array}$$
..............................

3)
$$\begin{array}{r} 1\;398 \\ -\;1\;005 \\ \hline \end{array}$$
..............................

4)
$$\begin{array}{r} 1\;381 \\ -\;1\;373 \\ \hline \end{array}$$
..............................

5)
$$\begin{array}{r} 2\;006 \\ -\;1\;662 \\ \hline \end{array}$$
..............................

6)
$$\begin{array}{r} 2\;620 \\ -\;2\;081 \\ \hline \end{array}$$
..............................

7)
$$\begin{array}{r} 1\;369 \\ -\;1\;271 \\ \hline \end{array}$$
..............................

8)
$$\begin{array}{r} 1\;953 \\ -\;1\;663 \\ \hline \end{array}$$
..............................

9)
$$\begin{array}{r} 2\;930 \\ -\;2\;400 \\ \hline \end{array}$$
..............................

10)
$$\begin{array}{r} 1\;852 \\ -\;1\;338 \\ \hline \end{array}$$
..............................

1)

  2034
− 1825

2)

  2920
− 1857

3)

  2313
− 1023

4)

  1810
− 1789

5)

  1575
− 1549

6)

  2366
− 1775

7)

  2726
− 2439

8)

  1012
− 1010

9)

  1149
− 1006

10)

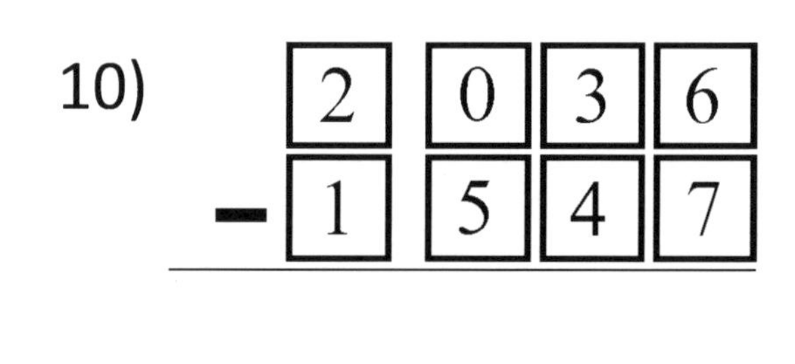

1) 1977 − 1172 = ..............................

2) 2894 − 2234 = ..............................

3) 1283 − 1003 = ..............................

4) 2414 − 2043 = ..............................

5) 1156 − 1072 = ..............................

6) 1580 − 1574 = ..............................

7) 1876 − 1413 = ..............................

8) 2427 − 1579 = ..............................

9) 2387 − 2194 = ..............................

10) 1758 − 1583 = ..............................

1)

$$\begin{array}{r} 1\ 2\ 5\ 6 \\ -\ 1\ 1\ 2\ 2 \\ \hline \end{array}$$

..................................................

2)

$$\begin{array}{r} 1\ 5\ 7\ 7 \\ -\ 1\ 1\ 5\ 7 \\ \hline \end{array}$$

..................................................

3)

$$\begin{array}{r} 1\ 9\ 3\ 2 \\ -\ 1\ 4\ 7\ 3 \\ \hline \end{array}$$

..................................................

4)

$$\begin{array}{r} 2\ 2\ 4\ 0 \\ -\ 1\ 2\ 6\ 2 \\ \hline \end{array}$$

..................................................

5)

$$\begin{array}{r} 2\ 3\ 8\ 4 \\ -\ 2\ 3\ 7\ 6 \\ \hline \end{array}$$

..................................................

6)

$$\begin{array}{r} 1\ 2\ 3\ 1 \\ -\ 1\ 0\ 7\ 8 \\ \hline \end{array}$$

..................................................

7)

$$\begin{array}{r} 2\ 9\ 2\ 2 \\ -\ 1\ 1\ 1\ 3 \\ \hline \end{array}$$

..................................................

8)

$$\begin{array}{r} 1\ 4\ 7\ 5 \\ -\ 1\ 1\ 4\ 3 \\ \hline \end{array}$$

..................................................

9)

$$\begin{array}{r} 2\ 1\ 9\ 4 \\ -\ 1\ 1\ 8\ 6 \\ \hline \end{array}$$

..................................................

10)

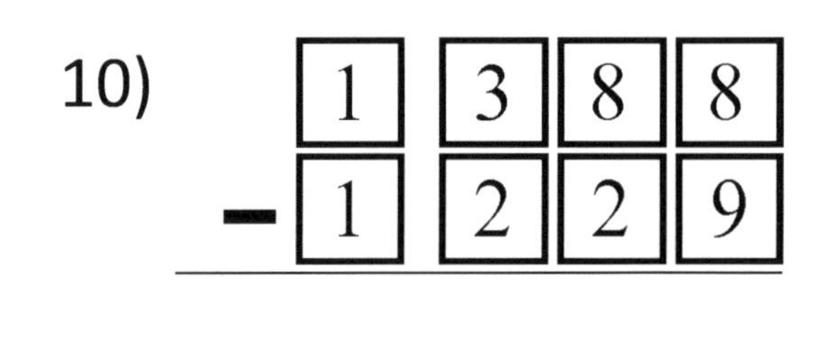

$$\begin{array}{r} 1\ 3\ 8\ 8 \\ -\ 1\ 2\ 2\ 9 \\ \hline \end{array}$$

..................................................

1) $\begin{array}{r} 2680 \\ -\ 2100 \\ \hline \end{array}$

..............................

2) $\begin{array}{r} 1573 \\ -\ 1176 \\ \hline \end{array}$

..............................

3) $\begin{array}{r} 1662 \\ -\ 1538 \\ \hline \end{array}$

..............................

4) $\begin{array}{r} 2404 \\ -\ 1232 \\ \hline \end{array}$

..............................

5) $\begin{array}{r} 1011 \\ -\ 1002 \\ \hline \end{array}$

..............................

6) $\begin{array}{r} 2571 \\ -\ 1757 \\ \hline \end{array}$

..............................

7) $\begin{array}{r} 1338 \\ -\ 1130 \\ \hline \end{array}$

..............................

8) $\begin{array}{r} 2309 \\ -\ 2012 \\ \hline \end{array}$

..............................

9) $\begin{array}{r} 2049 \\ -\ 1459 \\ \hline \end{array}$

..............................

10) 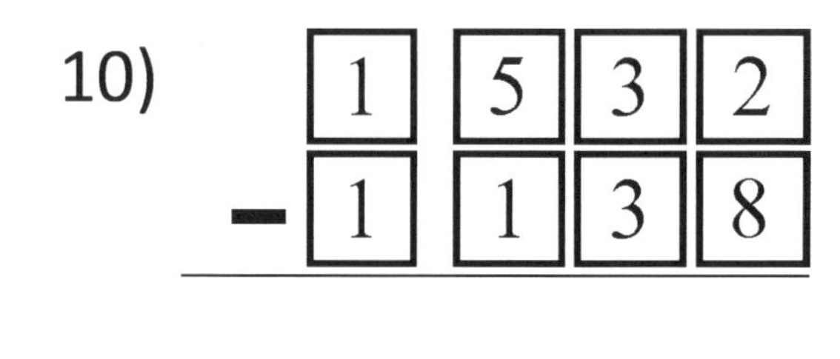

$\begin{array}{r} 1532 \\ -\ 1138 \\ \hline \end{array}$

..............................

1) 2 654
1 292

2)
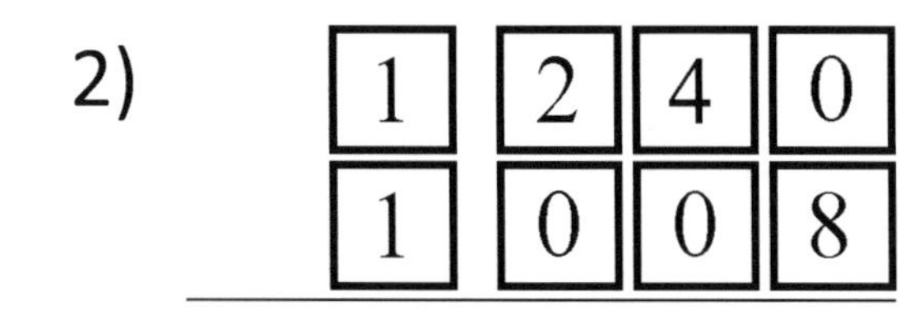


3) 1 308
1 142

4)
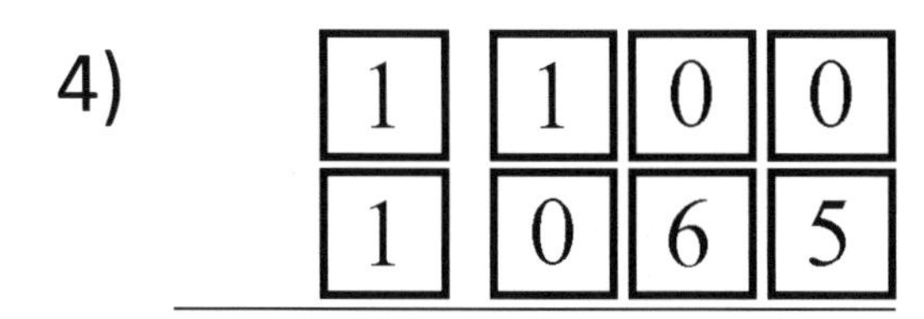


5) 2 455
1 618

6)
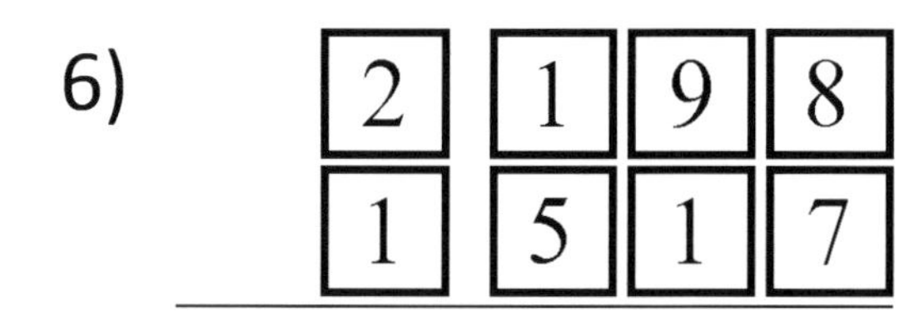


7) 1 301
1 297

8) 2 608
2 142

9) 2 719
1 282

10) 1 525
1 070

1) $2769 - 1584$

..............................

2) $2596 - 1866$

..............................

3) $1759 - 1462$

..............................

4) $1724 - 1098$

..............................

5) $1461 - 1089$

..............................

6) $2445 - 2259$

..............................

7) $2420 - 2195$

..............................

8) $2322 - 1600$

..............................

9) $2731 - 2507$

..............................

10)

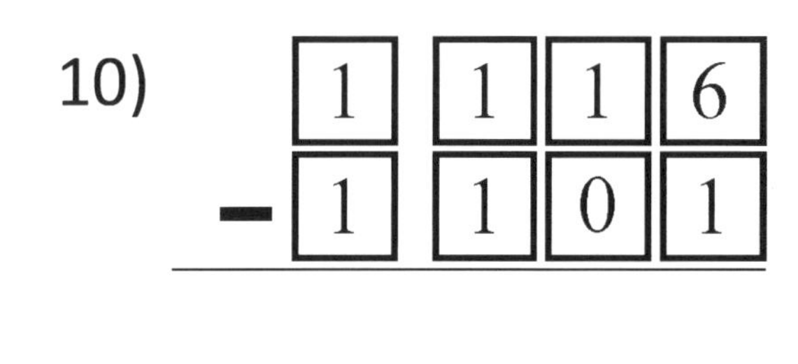


..............................

1) $2327 - 1830 =$ ..........

2) $2848 - 1918 =$ ..........

3) $2080 - 1290 =$ ..........

4) $2268 - 1770 =$ ..........

5) $2217 - 1238 =$ ..........

6) $1697 - 1504 =$ ..........

7) $2867 - 1128 =$ ..........

8) $1820 - 1499 =$ ..........

9) $2065 - 1358 =$ ..........

10) $1341 - 1109 =$ ..........

1)
  2 2 1 8
− 1 0 9 0

.............................

2)
  2 7 5 3
− 2 5 7 9

.............................

3)
  2 0 6 0
− 1 1 0 3

.............................

4)
  2 8 9 4
− 1 4 0 4

.............................

5)
  1 6 0 3
− 1 2 3 7

.............................

6)
  2 6 4 0
− 1 4 2 3

.............................

7)
  1 5 2 5
− 1 4 5 3

.............................

8)
  2 0 3 8
− 1 5 3 6

.............................

9)
  2 6 6 0
− 1 0 6 8

.............................

10)

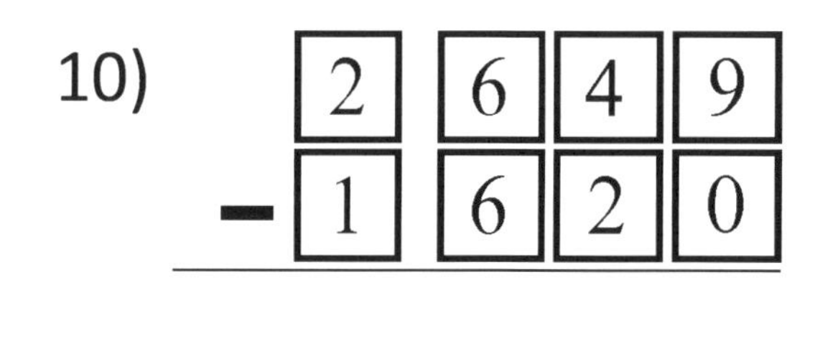


.............................

1) $\begin{array}{r} 1\ 0\ 9\ 6 \\ -\ 1\ 0\ 0\ 7 \\ \hline \end{array}$

..........................................................

2) $\begin{array}{r} 1\ 0\ 9\ 8 \\ -\ 1\ 0\ 8\ 0 \\ \hline \end{array}$

..........................................................

3) $\begin{array}{r} 2\ 1\ 7\ 3 \\ -\ 1\ 3\ 2\ 2 \\ \hline \end{array}$

..........................................................

4) $\begin{array}{r} 2\ 6\ 9\ 5 \\ -\ 1\ 0\ 5\ 6 \\ \hline \end{array}$

..........................................................

5) $\begin{array}{r} 2\ 0\ 7\ 7 \\ -\ 1\ 3\ 3\ 7 \\ \hline \end{array}$

..........................................................

6) $\begin{array}{r} 2\ 3\ 2\ 2 \\ -\ 1\ 4\ 2\ 2 \\ \hline \end{array}$

..........................................................

7) $\begin{array}{r} 1\ 0\ 6\ 9 \\ -\ 1\ 0\ 5\ 4 \\ \hline \end{array}$

..........................................................

8) $\begin{array}{r} 1\ 9\ 9\ 9 \\ -\ 1\ 7\ 0\ 0 \\ \hline \end{array}$

..........................................................

9) $\begin{array}{r} 2\ 3\ 2\ 8 \\ -\ 1\ 6\ 6\ 9 \\ \hline \end{array}$

..........................................................

10)

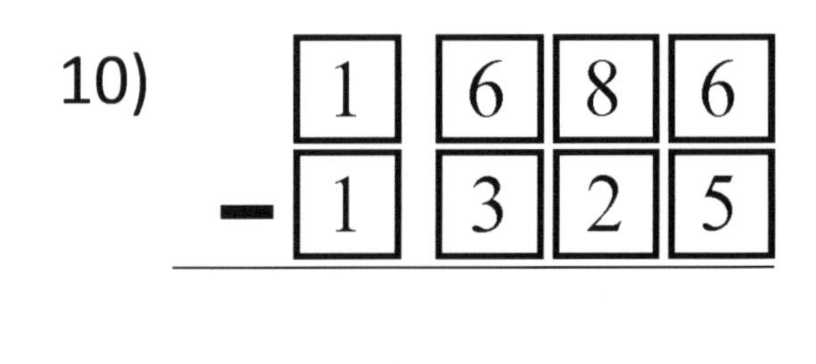


..........................................................

# ANSWERS

Page 1

1) $234 + 175 = 409$
2) $229 + 130 = 359$
3) $271 + 119 = 390$
4) $147 + 259 = 406$
5) $125 + 266 = 391$
6) $129 + 146 = 275$
7) $249 + 191 = 440$
8) $277 + 117 = 394$
9) $154 + 266 = 420$
10) $285 + 240 = 525$
11) $286 + 235 = 521$
12) $119 + 247 = 366$
13) $261 + 251 = 512$
14) $292 + 157 = 449$
15) $289 + 245 = 534$

Page 2

1) $125 + 279 = 404$
2) $164 + 293 = 457$
3) $222 + 161 = 383$
4) $130 + 278 = 408$
5) $123 + 165 = 288$
6) $299 + 156 = 455$
7) $282 + 169 = 451$
8) $120 + 127 = 247$
9) $289 + 271 = 560$
10) $180 + 102 = 282$
11) $173 + 121 = 294$
12) $162 + 252 = 414$
13) $128 + 107 = 235$
14) $199 + 262 = 461$
15) $183 + 290 = 473$

Page 3

1) $193 + 127 = 320$
2) $187 + 283 = 470$
3) $174 + 261 = 435$
4) $223 + 186 = 409$
5) $217 + 106 = 323$
6) $104 + 217 = 321$
7) $212 + 265 = 477$
8) $215 + 262 = 477$
9) $167 + 194 = 361$
10) $133 + 273 = 406$
11) $130 + 256 = 386$
12) $215 + 233 = 448$
13) $148 + 200 = 348$
14) $225 + 204 = 429$
15) $295 + 167 = 462$

Page 4

1) $155 + 115 = 270$
2) $199 + 252 = 451$
3) $223 + 167 = 390$
4) $184 + 269 = 453$
5) $150 + 210 = 360$
6) $238 + 212 = 450$
7) $254 + 235 = 489$
8) $274 + 135 = 409$
9) $119 + 190 = 309$
10) $149 + 213 = 362$
11) $286 + 177 = 463$
12) $200 + 130 = 330$
13) $234 + 250 = 484$
14) $188 + 185 = 373$
15) $104 + 104 = 208$

## Page 5

1) $168 + 158 = 326$
2) $144 + 136 = 280$
3) $294 + 250 = 544$
4) $293 + 235 = 528$
5) $249 + 246 = 495$
6) $278 + 215 = 493$
7) $170 + 118 = 288$
8) $273 + 278 = 551$
9) $299 + 265 = 564$
10) $146 + 263 = 409$
11) $266 + 159 = 425$
12) $253 + 241 = 494$
13) $189 + 109 = 298$
14) $167 + 118 = 285$
15) $149 + 203 = 352$

## Page 6

1) $567 + 627 = 1194$
2) $449 + 585 = 1034$
3) $213 + 525 = 738$
4) $768 + 538 = 1306$
5) $439 + 108 = 547$
6) $736 + 673 = 1409$
7) $367 + 420 = 787$
8) $178 + 531 = 709$
9) $422 + 469 = 891$
10) $394 + 715 = 1109$
11) $765 + 386 = 1151$
12) $406 + 535 = 941$
13) $431 + 147 = 578$
14) $796 + 559 = 1355$
15) $461 + 543 = 1004$

## Page 7

1) $203 + 697 = 900$
2) $237 + 795 = 1032$
3) $469 + 397 = 866$
4) $766 + 701 = 1467$
5) $190 + 374 = 564$
6) $281 + 320 = 601$
7) $714 + 254 = 968$
8) $166 + 530 = 696$
9) $309 + 178 = 487$
10) $695 + 252 = 947$
11) $213 + 756 = 969$
12) $512 + 422 = 934$
13) $182 + 683 = 865$
14) $699 + 746 = 1445$
15) $262 + 162 = 424$

## Page 8

1) $576 + 734 = 1310$
2) $369 + 458 = 827$
3) $503 + 444 = 947$
4) $498 + 696 = 1194$
5) $273 + 380 = 653$
6) $247 + 285 = 532$
7) $615 + 713 = 1328$
8) $587 + 498 = 1085$
9) $120 + 539 = 659$
10) $128 + 185 = 313$
11) $486 + 339 = 825$
12) $673 + 726 = 1399$
13) $716 + 672 = 1388$
14) $545 + 186 = 731$
15) $549 + 361 = 910$

## Page 9

1) $764 + 143 = 907$
2) $651 + 305 = 956$
3) $548 + 670 = 1218$
4) $284 + 599 = 883$
5) $295 + 231 = 526$
6) $337 + 161 = 498$
7) $443 + 542 = 985$
8) $631 + 700 = 1331$
9) $412 + 247 = 659$
10) $712 + 189 = 901$
11) $118 + 154 = 272$
12) $110 + 407 = 517$
13) $638 + 168 = 806$
14) $531 + 779 = 1310$
15) $411 + 395 = 806$

## Page 10

1) $688 + 788 = 1476$
2) $647 + 728 = 1375$
3) $618 + 278 = 896$
4) $552 + 697 = 1249$
5) $472 + 163 = 635$
6) $710 + 134 = 844$
7) $344 + 467 = 811$
8) $480 + 239 = 719$
9) $304 + 288 = 592$
10) $701 + 443 = 1144$
11) $607 + 750 = 1357$
12) $597 + 590 = 1187$
13) $102 + 106 = 208$
14) $531 + 246 = 777$
15) $321 + 182 = 503$

## Page 11

1) $2240 + 1931 = 4171$
2) $1100 + 2855 = 3955$
3) $1866 + 1851 = 3717$
4) $1871 + 1351 = 3222$
5) $1705 + 2518 = 4223$
6) $1681 + 2965 = 4646$
7) $2396 + 2198 = 4594$
8) $1708 + 1460 = 3168$
9) $2114 + 1887 = 4001$
10) $1615 + 1893 = 3508$

## Page 12

1) $1655 + 2766 = 4421$
2) $2170 + 2092 = 4262$
3) $1893 + 1274 = 3167$
4) $1807 + 1872 = 3679$
5) $2718 + 1442 = 4160$
6) $2286 + 1450 = 3736$
7) $1679 + 1252 = 2931$
8) $2381 + 1641 = 4022$
9) $1197 + 2125 = 3322$
10) $1361 + 2054 = 3415$

### Page 13

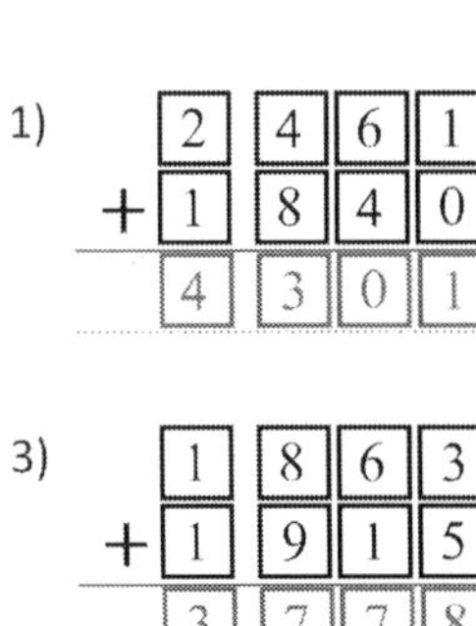

1) 2461 + 1840 = 4301

2) 2397 + 2085 = 4482

3) 1863 + 1915 = 3778

4) 1796 + 1304 = 3100

5) 1305 + 1434 = 2739

6) 2654 + 1241 = 3895

7) 2169 + 1173 = 3342

8) 2608 + 1380 = 3988

9) 1132 + 1105 = 2237

10) 1183 + 1748 = 2931

### Page 14

1) 2701 + 1271 = 3972

2) 1407 + 1801 = 3208

3) 1981 + 1446 = 3427

4) 2053 + 2595 = 4648

5) 2389 + 1466 = 3855

6) 2846 + 2893 = 5739

7) 1820 + 2927 = 4747

8) 2400 + 2718 = 5118

9) 2141 + 2633 = 4774

10) 2013 + 2700 = 4713

### Page 15

1) 1954 + 1247 = 3201

2) 2291 + 2841 = 5132

3) 1733 + 2382 = 4115

4) 2834 + 1368 = 4202

5) 2192 + 2161 = 4353

6) 2327 + 1537 = 3864

7) 1715 + 2717 = 4432

8) 1902 + 2859 = 4761

9) 2445 + 2743 = 5188

10) 1499 + 2943 = 4442

### Page 16

1) 2277 + 2555 = 4832

2) 1171 + 1871 = 3042

3) 2512 + 2599 = 5111

4) 1602 + 1979 = 3581

5) 2194 + 1018 = 3212

6) 2884 + 1772 = 4656

7) 1441 + 2852 = 4293

8) 1373 + 2760 = 4133

9) 2753 + 1635 = 4388

10) 1816 + 2307 = 4123

Page 17

1) 2463 + 2145 = 4608

2) 1733 + 2407 = 4140

3) 2884 + 2773 = 5657

4) 1366 + 2844 = 4210

5) 1204 + 2303 = 3507

6) 2257 + 1466 = 3723

7) 1548 + 1481 = 3029

8) 1060 + 1432 = 2492

9) 2887 + 2683 = 5570

10) 1940 + 1408 = 3348

Page 18

1) 1887 + 1623 = 3510

2) 2306 + 2693 = 4999

3) 1555 + 1738 = 3293

4) 1040 + 2034 = 3074

5) 2968 + 1600 = 4568

6) 2394 + 2909 = 5303

7) 1431 + 2513 = 3944

8) 1050 + 2285 = 3335

9) 1879 + 1517 = 3396

10) 2420 + 1881 = 4301

Page 19

1) 2216 + 2674 = 4890

2) 1286 + 2318 = 3604

3) 1416 + 1634 = 3050

4) 2160 + 1856 = 4016

5) 2926 + 2321 = 5247

6) 1393 + 1853 = 3246

7) 2716 + 1340 = 4056

8) 2014 + 2569 = 4583

9) 1853 + 1360 = 3213

10) 1279 + 1750 = 3029

Page 20

1) 2466 + 2635 = 5101

2) 2662 + 2933 = 5595

3) 2384 + 2720 = 5104

4) 2623 + 2113 = 4736

5) 2097 + 2416 = 4513

6) 1125 + 1023 = 2148

7) 1591 + 1099 = 2690

8) 1488 + 2886 = 4374

9) 2313 + 1699 = 4012

10) 1252 + 1454 = 2706

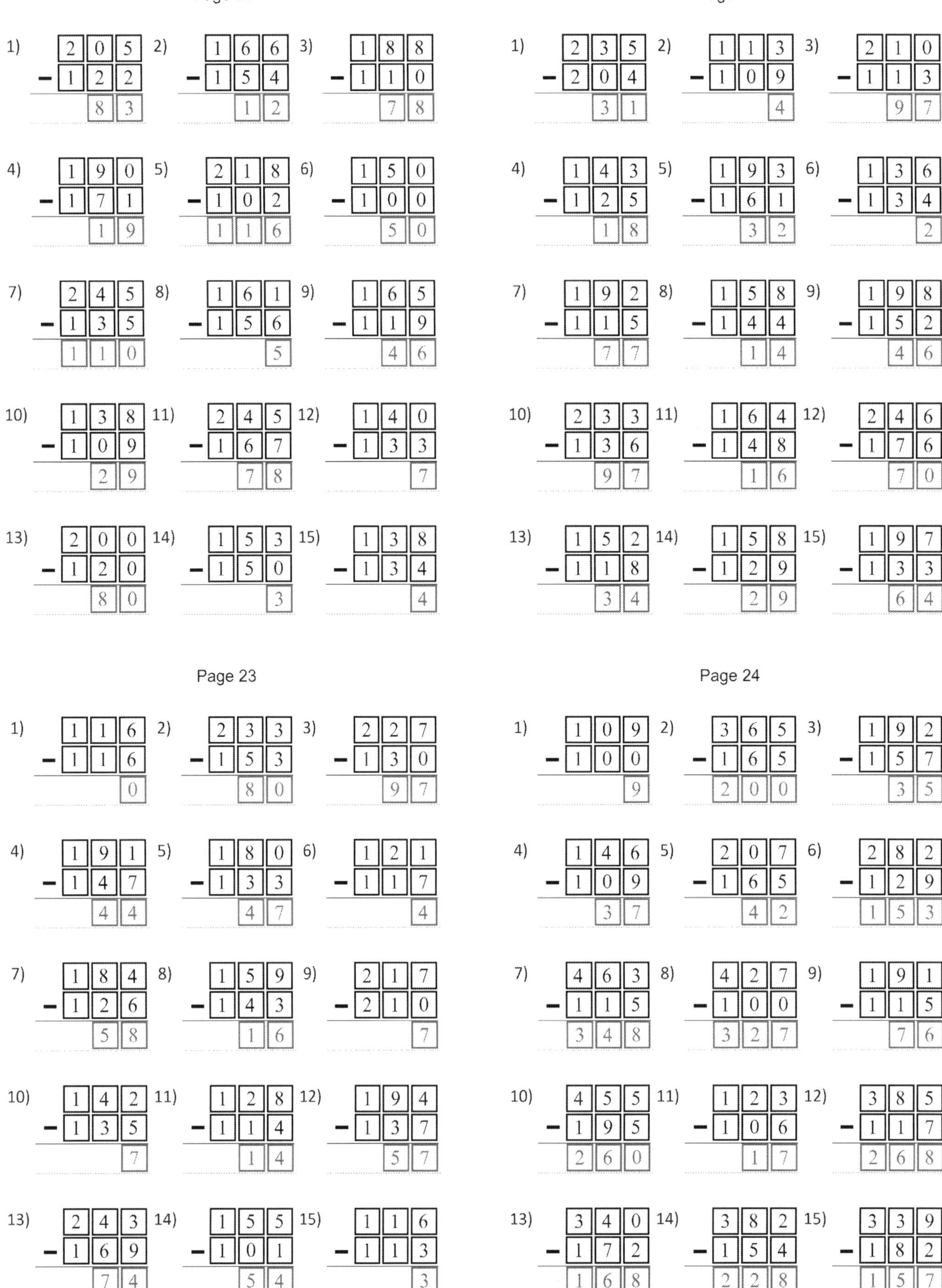
Page 21
1) 205 - 122 83
2) 166 - 154 12
3) 188 - 110 78
4) 190 - 171 19
5) 218 - 102 116
6) 150 - 100 50
7) 245 - 135 110
8) 161 - 156 5
9) 165 - 119 46
10) 138 - 109 29
11) 245 - 167 78
12) 140 - 133 7
13) 200 - 120 80
14) 153 - 150 3
15) 138 - 134 4
Page 22
1) 235 - 204 31
2) 113 - 109 4
3) 210 - 113 97
4) 143 - 125 18
5) 193 - 161 32
6) 136 - 134 2
7) 192 - 115 77
8) 158 - 144 14
9) 198 - 152 46
10) 233 - 136 97
11) 164 - 148 16
12) 246 - 176 70
13) 152 - 118 34
14) 158 - 129 29
15) 197 - 133 64
Page 23
1) 116 - 116 0
2) 233 - 153 80
3) 227 - 130 97
4) 191 - 147 44
5) 180 - 133 47
6) 121 - 117 4
7) 184 - 126 58
8) 159 - 143 16
9) 217 - 210 7
10) 142 - 135 7
11) 128 - 114 14
12) 194 - 137 57
13) 243 - 169 74
14) 155 - 101 54
15) 116 - 113 3
Page 24
1) 109 - 100 9
2) 365 - 165 200
3) 192 - 157 35
4) 146 - 109 37
5) 207 - 165 42
6) 282 - 129 153
7) 463 - 115 348
8) 427 - 100 327
9) 191 - 115 76
10) 455 - 195 260
11) 123 - 106 17
12) 385 - 117 268
13) 340 - 172 168
14) 382 - 154 228
15) 339 - 182 157

## Page 25

1) 331 - 172 = 159
2) 417 - 135 = 282
3) 152 - 127 = 25
4) 344 - 135 = 209
5) 148 - 127 = 21
6) 397 - 145 = 252
7) 319 - 225 = 94
8) 399 - 237 = 162
9) 401 - 190 = 211
10) 389 - 155 = 234
11) 261 - 163 = 98
12) 280 - 114 = 166
13) 340 - 248 = 92
14) 308 - 133 = 175
15) 454 - 113 = 341

## Page 26

1) 468 - 170 = 298
2) 449 - 130 = 319
3) 135 - 128 = 7
4) 408 - 111 = 297
5) 420 - 171 = 249
6) 162 - 102 = 60
7) 381 - 195 = 186
8) 159 - 100 = 59
9) 257 - 152 = 105
10) 114 - 112 = 2
11) 204 - 179 = 25
12) 222 - 199 = 23
13) 399 - 100 = 299
14) 351 - 120 = 231
15) 381 - 212 = 169

## Page 27

1) 488 - 131 = 357
2) 295 - 148 = 147
3) 106 - 101 = 5
4) 289 - 192 = 97
5) 113 - 100 = 13
6) 165 - 111 = 54
7) 124 - 111 = 13
8) 135 - 106 = 29
9) 355 - 156 = 199
10) 122 - 120 = 2
11) 253 - 172 = 81
12) 450 - 243 = 207
13) 353 - 153 = 200
14) 251 - 182 = 69
15) 398 - 242 = 156

## Page 28

1) 332 - 317 = 15
2) 438 - 275 = 163
3) 263 - 174 = 89
4) 241 - 178 = 63
5) 263 - 222 = 41
6) 242 - 173 = 69
7) 672 - 177 = 495
8) 592 - 151 = 441
9) 128 - 100 = 28
10) 359 - 238 = 121
11) 702 - 595 = 107
12) 489 - 326 = 163
13) 649 - 513 = 136
14) 663 - 243 = 420
15) 508 - 229 = 279

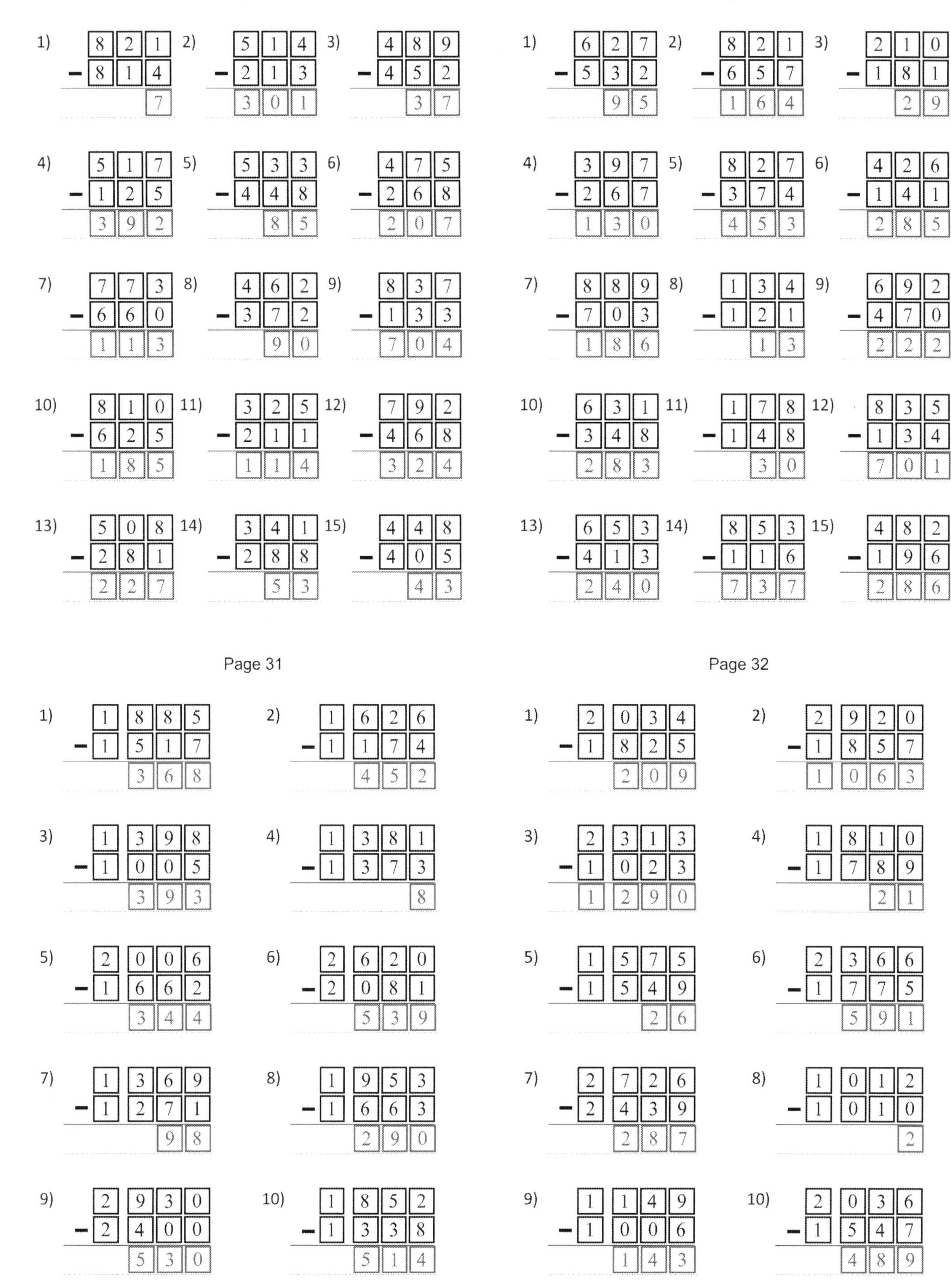
Page 29
1) 821 − 814 = 7
2) 514 − 213 = 301
3) 489 − 452 = 37
4) 517 − 125 = 392
5) 533 − 448 = 85
6) 475 − 268 = 207
7) 773 − 660 = 113
8) 462 − 372 = 90
9) 837 − 133 = 704
10) 810 − 625 = 185
11) 325 − 211 = 114
12) 792 − 468 = 324
13) 508 − 281 = 227
14) 341 − 288 = 53
15) 448 − 405 = 43
Page 30
1) 627 − 532 = 95
2) 821 − 657 = 164
3) 210 − 181 = 29
4) 397 − 267 = 130
5) 827 − 374 = 453
6) 426 − 141 = 285
7) 889 − 703 = 186
8) 134 − 121 = 13
9) 692 − 470 = 222
10) 631 − 348 = 283
11) 178 − 148 = 30
12) 835 − 134 = 701
13) 653 − 413 = 240
14) 853 − 116 = 737
15) 482 − 196 = 286
Page 31
1) 1885 − 1517 = 368
2) 1626 − 1174 = 452
3) 1398 − 1005 = 393
4) 1381 − 1373 = 8
5) 2006 − 1662 = 344
6) 2620 − 2081 = 539
7) 1369 − 1271 = 98
8) 1953 − 1663 = 290
9) 2930 − 2400 = 530
10) 1852 − 1338 = 514
Page 32
1) 2034 − 1825 = 209
2) 2920 − 1857 = 1063
3) 2313 − 1023 = 1290
4) 1810 − 1789 = 21
5) 1575 − 1549 = 26
6) 2366 − 1775 = 591
7) 2726 − 2439 = 287
8) 1012 − 1010 = 2
9) 1149 − 1006 = 143
10) 2036 − 1547 = 489

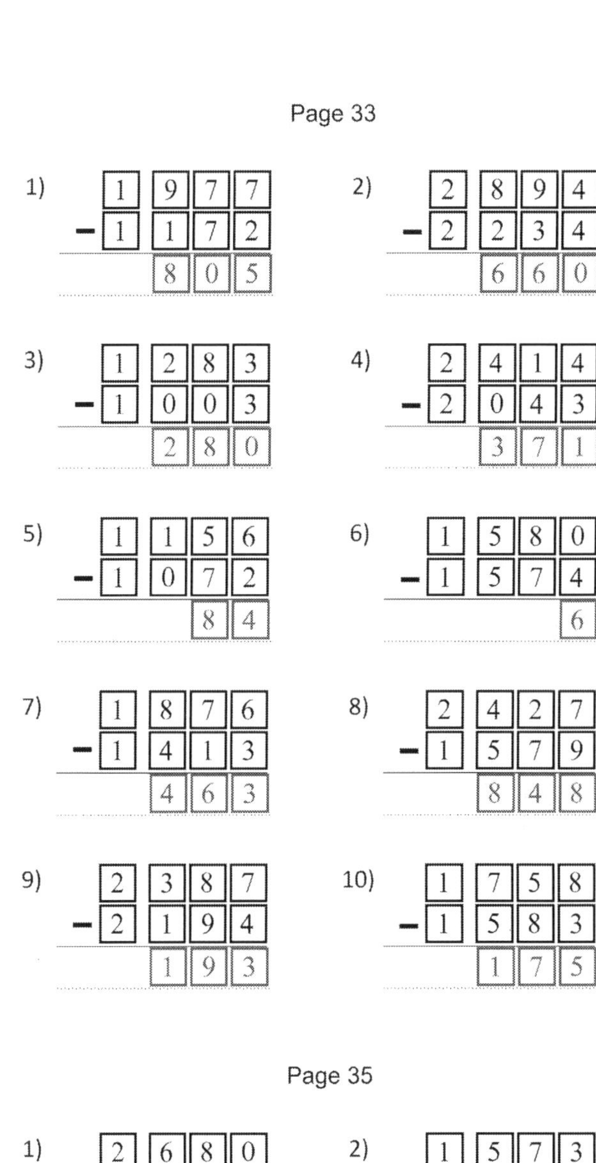

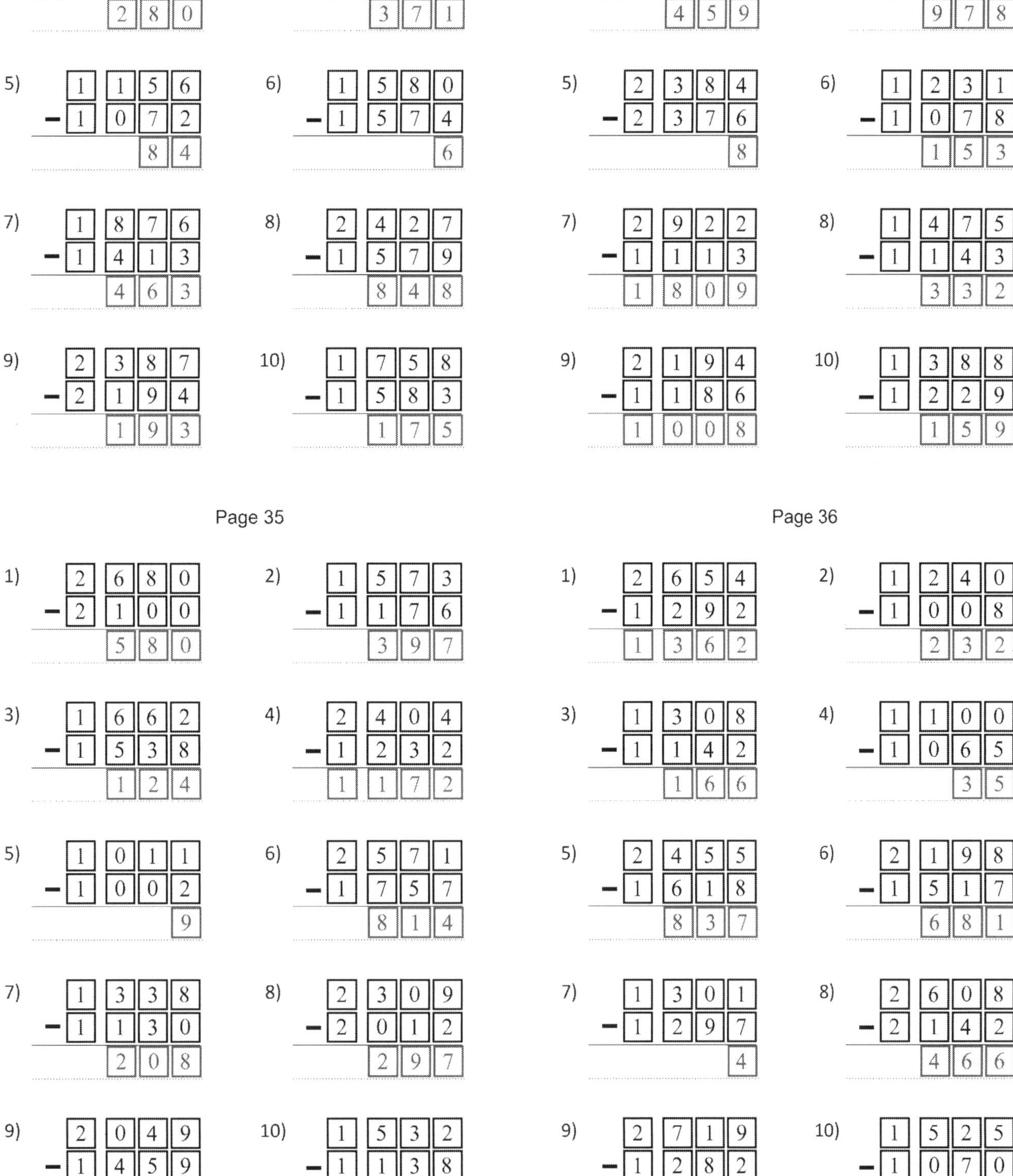

## Page 33

1) 1977 − 1172 = 805
2) 2894 − 2234 = 660
3) 1283 − 1003 = 280
4) 2414 − 2043 = 371
5) 1156 − 1072 = 84
6) 1580 − 1574 = 6
7) 1876 − 1413 = 463
8) 2427 − 1579 = 848
9) 2387 − 2194 = 193
10) 1758 − 1583 = 175

## Page 34

1) 1256 − 1122 = 134
2) 1577 − 1157 = 420
3) 1932 − 1473 = 459
4) 2240 − 1262 = 978
5) 2384 − 2376 = 8
6) 1231 − 1078 = 153
7) 2922 − 1113 = 1809
8) 1475 − 1143 = 332
9) 2194 − 1186 = 1008
10) 1388 − 1229 = 159

## Page 35

1) 2680 − 2100 = 580
2) 1573 − 1176 = 397
3) 1662 − 1538 = 124
4) 2404 − 1232 = 1172
5) 1011 − 1002 = 9
6) 2571 − 1757 = 814
7) 1338 − 1130 = 208
8) 2309 − 2012 = 297
9) 2049 − 1459 = 590
10) 1532 − 1138 = 394

## Page 36

1) 2654 − 1292 = 1362
2) 1240 − 1008 = 232
3) 1308 − 1142 = 166
4) 1100 − 1065 = 35
5) 2455 − 1618 = 837
6) 2198 − 1517 = 681
7) 1301 − 1297 = 4
8) 2608 − 2142 = 466
9) 2719 − 1282 = 1437
10) 1525 − 1070 = 455

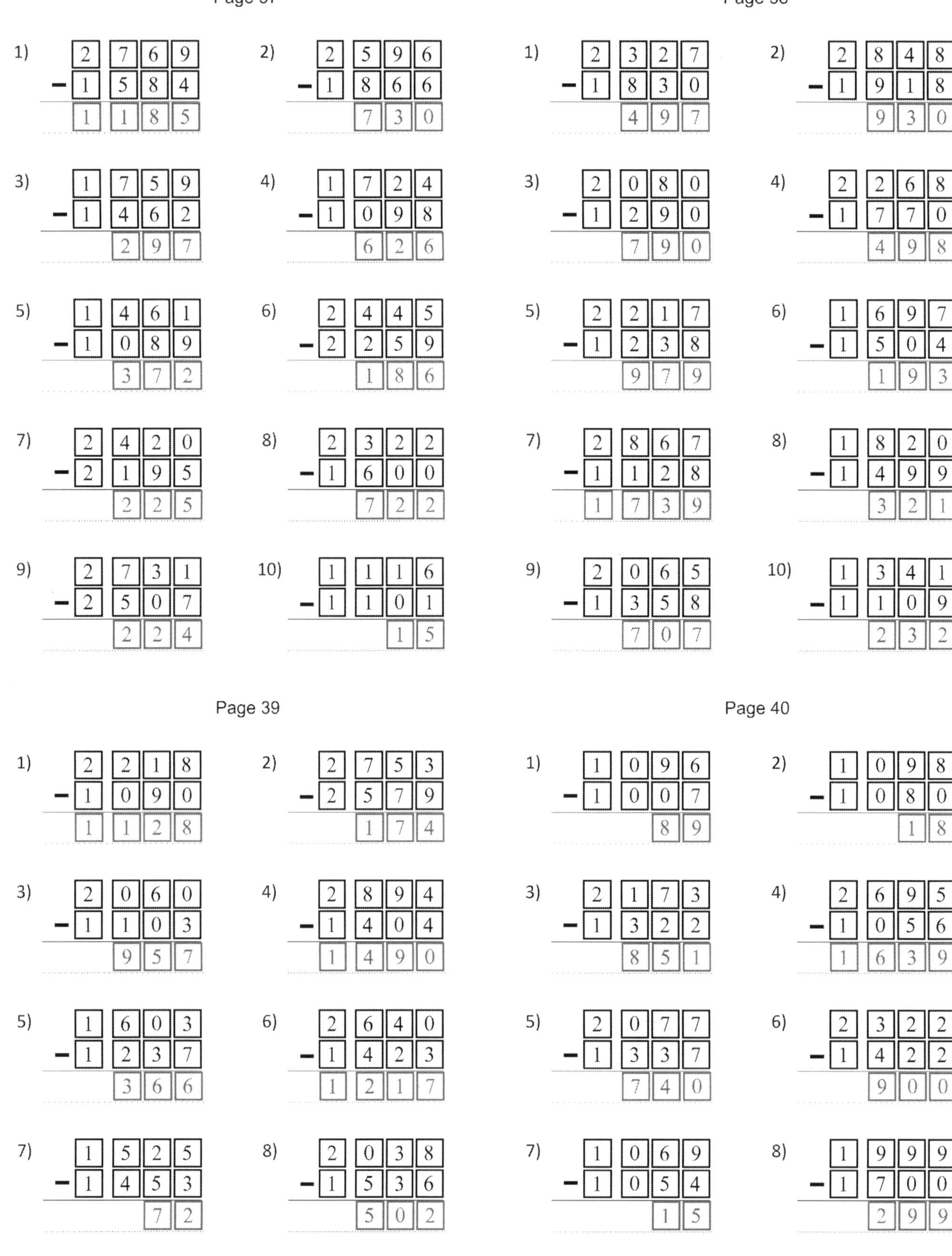

Page 37

1) 2769 − 1584 = 1185
2) 2596 − 1866 = 730
3) 1759 − 1462 = 297
4) 1724 − 1098 = 626
5) 1461 − 1089 = 372
6) 2445 − 2259 = 186
7) 2420 − 2195 = 225
8) 2322 − 1600 = 722
9) 2731 − 2507 = 224
10) 1116 − 1101 = 15

Page 38

1) 2327 − 1830 = 497
2) 2848 − 1918 = 930
3) 2080 − 1290 = 790
4) 2268 − 1770 = 498
5) 2217 − 1238 = 979
6) 1697 − 1504 = 193
7) 2867 − 1128 = 1739
8) 1820 − 1499 = 321
9) 2065 − 1358 = 707
10) 1341 − 1109 = 232

Page 39

1) 2218 − 1090 = 1128
2) 2753 − 2579 = 174
3) 2060 − 1103 = 957
4) 2894 − 1404 = 1490
5) 1603 − 1237 = 366
6) 2640 − 1423 = 1217
7) 1525 − 1453 = 72
8) 2038 − 1536 = 502
9) 2660 − 1068 = 1592
10) 2649 − 1620 = 1029

Page 40

1) 1096 − 1007 = 89
2) 1098 − 1080 = 18
3) 2173 − 1322 = 851
4) 2695 − 1056 = 1639
5) 2077 − 1337 = 740
6) 2322 − 1422 = 900
7) 1069 − 1054 = 15
8) 1999 − 1700 = 299
9) 2328 − 1669 = 659
10) 1686 − 1325 = 361

Made in the USA
Las Vegas, NV
31 August 2022

54443525R00033